Animal Features

Mary Ann Hoffman

New York

This is a fish.
It has fins.

Animal Features

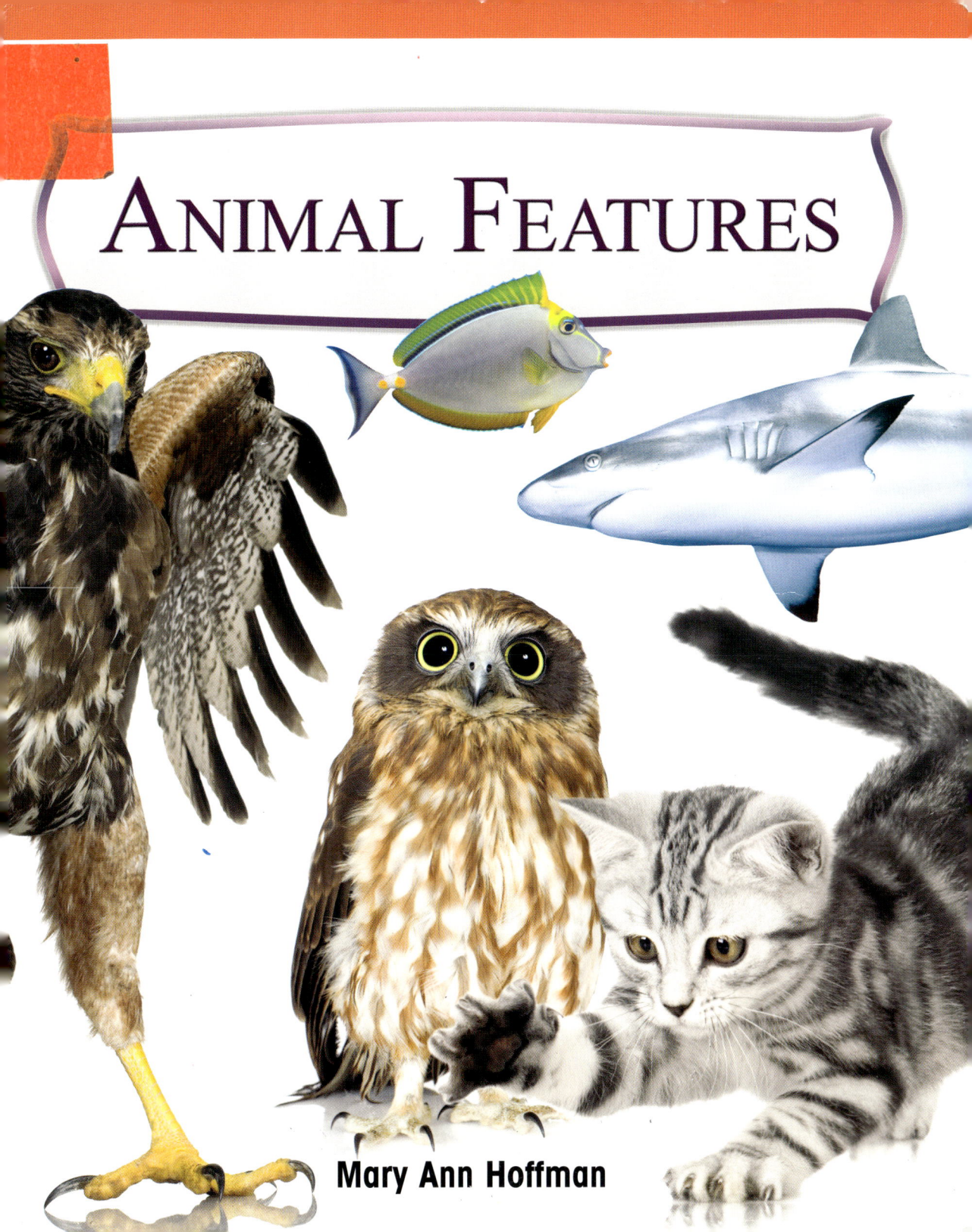

Mary Ann Hoffman

Published in 2009 by The Rosen Publishing Group, Inc.
29 East 21st Street, New York, NY 10010

Copyright © 2009 by The Rosen Publishing Group, Inc.

All rights reserved. No part of this book may be reproduced in any form without permission in writing from the publisher, except by a reviewer.

Book Design: Michael J. Flynn

Photo Credits: Cover (tropical fish), pp. 2, 7 (tropical fish) © TatjanaRittner/Shutterstock; cover (cat, hawk, owl), pp. 3, 4, 5, 7 (cat, hawk, owl) © Eric Isselée/Shutterstock; cover (shark), pp. 6, 7 (shark) © cbpix/Shutterstock.

ISBN: 978-1-4042-7967-4
6-pack ISBN: 978-1-4042-7968-1

Manufactured in the United States of America

CPSIA Compliance Information: Batch #WR212070RC: For Further Information contact Rosen Publishing, New York, New York at 1-800-237-9932

Word Count: 37

This is a hawk.
It has wings.

This is a cat.
It has claws.

This is an owl.
It has a beak.

This is a shark.
It has big teeth!

Animal Features

	beak	claws	fins	teeth	wings
fish			X		
hawk	X	X			X
cat		X		X	
owl	X	X			X
shark			X	X	

7

Words to Know

beak

claws

fins

teeth

wings